An Altitude
SuperGuide

The
Columbia
Icefield

An Altitude SuperGuide

The Columbia Icefield

Robert Sandford

Altitude Publishing
Banff Alberta Canada

Columbia Icefield with Mt. Athabasca on the left
(Front cover)

Mountain Goat
(Front cover inset left)

Two bighorn sheep
(Front cover inset right)

Aerial View of the Athabasca Glacier
(Page 2) This photograph shows many of the prominent features that visitors see from the Icefields Parkway, including the Columbia Icefield proper as it gives rise to the three icefalls that form the upper part of the Athabasca Glacier. You can also see the Snowcoach Road on the left side of the ice approaching the lowest icefall and the knife-edged lateral moraine on the left of the outwash, past which the glacier has already receded. Sunwapta Lake is also visible at the glacier's snout. The prominent white streak at the right of the photograph is Dome Glacier.

Climbing in the Columbia Icefield
(Opposite) Demanding peaks and breathtaking vistas have made the Columbia Icefield area a long-time favourite of mountain climbers.

The topographic map is based on information taken from the National Topographic System map sheet numbers MCR 221, Jasper National Park © 1985. Her Majesty the Queen in Right of Canada with permission of Energy, Mines and Resources Canada.

Extreme care has been taken to ensure that all information in this book is accurate and up to date, but neither the author nor the publisher can be held legally responsible for any errors that may appear.

Canadian Cataloguing in Publication Data
Sandford, Robert W.
The Columbia Icefield
Includes bibliographical references.
ISBN 1-55153-619-6
(SuperGuide)
1. Columbia Icefield (B.C. and Alta.) - Guidebooks.
2. Athabasca Glacier (Alta.) - Guidebooks
I. Title
GB588.15.826 1992
551.3'12'09712332
C92-091443-8

9 8 7 6 5

www.altitudepublishing.com

Editor: Elizabeth Wilson
Design: Robert MacDonald, MediaClones Inc.
Cover design: Stephen Hutchings

All photographs in this book are by Robert Sanford, except for the ones on the following pages: Stephen Hutchings 21, 39 top, 42 top, back cover; Carole Harmon 19 top, 38 top; Nicholas Morant 59 top; Dennis Schmidt front cover inset right; Esther Schmidt front cover inset left. The photographs on pages 24 and 57 are reproduced courtesy of the Whyte Museum of the Canadian Rockies.

Made in Western Canada
Printed and bound in Western Canada by Friesen Printers, Altona Manitoba, using Canadian-made paper and vegetable-based inks.

We acknowledge the financial support of the Government of Canada through the Book Publishing Industry Development Program (BPIDP) for our publishing activities.

Altitude GreenTree Program
Altitude will plant in Canada twice as many trees as were used in the manufacturing of this book.

Dedication
For Walter Sandford, who led his young brother through the stone boneyards of the Earth's past, backwards into time.

Altitude Publishing Canada Ltd.

1500 Railway Ave
Canmore, Alberta
T1W 1P6

Contents

Welcome to the Columbia Icefield   6

Geology and the History of Glaciers   10

Visiting the Columbia Icefield   24

A Visual Tour of the Columbia Icefield   32

Life in the Columbia Icefield   48

For More Information   62

Glossary   63

Welcome to the Columbia Icefield

The Columbia Icefield Chalet
The Columbia Icefield Chalet was built by Brewster Transport in 1939. At that time the Athabasca Glacier was significantly larger than it is today.

The Columbia Icefield is an extraordinary window back into the time of the ice. In this amazing place, it is possible to experience what much of North America and Europe must have been like while they were in the throes of the last Ice Age. Here, it is actually possible to stand next to a glacier, and witness the forces that have shaped the surface of the planet as we know it today.

This book shows how the dramatic Columbia Icefield landscape was created and explains the dynamics of the glaciers that exist here. You'll find sections on the history of the area, its geology, its plants and animals as well as tales of the complex and sometimes tragic relationship between humans and glaciers. Near the back of the book, you'll find a glossary of terms and a list of other books on the area that you might enjoy reading. As captivating as the photographs you'll find throughout this book are, however, they don't compare to being there.

Geology and the History of Glaciers

Crevasses
Glacial fissures, or crevasses, can be beautiful as well as dangerous. Here, in a crevasse, two visitors explore part of a glacier from beneath its surface.

By any standard, the Columbia Icefield is an incredible geographical feature. It is a high basin of accumulated snow and ice that straddles 325 square kilometres (roughly 125 square miles), of the Great Divide that separates British Columbia from Alberta. Located 52° north latitude and 117° west longitude, the Columbia Icefield also straddles Banff and Jasper National Parks, and contributes significantly to their designations as United Nations World Heritage Sites. By definition, an icefield is simply an up-

land area of ice that feeds two or more glaciers.

Of the 22 highest peaks in the Canadian Rockies, 11 are found in or close to the Columbia Icefield. The highest mountain in Alberta, 3745 metre (12,283 foot) Mt. Columbia, juts out of the northern edge of this field of ice. Much of the basin itself is over 2600 metres (8528 feet) above sea level.

The Columbia Icefield is located just east of a major gap in the Columbia Mountains of British Columbia, making the Great Divide the first major obstacle to moisture-laden winds blowing eastward from the Pacific. The basin's high altitude, combined with the wind-blocking effect of the presence of high peaks, dramatically increases snowfall in the Columbia Icefield area. At least 10 metres (more than 30 feet) of snow fall on this basin every year, by far more snow than falls anywhere for hundreds of kilometres to the north or south. The snow and ice in this basin is, in many places, more than 300 metres, (1000 feet) deep. Steep cliffs of ice cap even the summits of the icefield mountains to a depth of as much as 100 metres (328 feet).

Out of this basin flow six major glaciers. Though by no means the largest one in the Icefield, the Athabasca Glacier is definitely the best known. It is also the only one accessible by road. The Athabasca Glacier presents itself each year to hundreds of thousands of travellers who choose the Icefield Parkway as their route from Lake Louise to Jasper. It is a classic outlet valley glacier and one of the most spectacular natural features in the Canadian Rockies.

Though the nearby but less accessible Saskatchewan Glacier is almost twice the Athabasca's size, the latter is still an impressive mass of ice. As of the spring of 1990, it was still six kilometres (four miles) long when measured from its upper edge to its toe. Its average width is one kilometre (0.6 miles). When measured from mid-glacier, the turn-around point for the snowcoaches that tour its surface, it is 300 metres

How Is a Glacier Formed?

Glaciers form in places where more snow accumulates annually than melts. When snow falls in the Rockies, individual flakes fall one upon the other, glistening and gradually deepening into the romantic image of the Canadian winter. As snow continues to fall and deepen, the sheer weight of accumulation changes the nature of the flakes. As pressure builds, the radial arms, outstretched and intertwined, break off. Eventually the aging flakes break down into a form of granular snow called hoar.

In most places in the world, the life of hoar snow comes to an end with the hot days of springtime – the aging snow dies back into water as it melts. But at the poles, and in the high places of the Earth's mountains, the snow that falls in winter doesn't always melt. It deepens, layer after layer, year after passing year. At a certain depth, 30 metres or so, the compressed snow slowly becomes ice. Under its own weight, and in response to dictates of its own crystalline nature, this ice moves. This is the ice of the eons – this is glacier ice.

Just How Big is 325 Square Kilometres?

The Columbia Icefield covers an area of roughly 325 square kilometres (125.5 square miles). The Athabasca Glacier is about 6 square kilometres (2.3 square miles) big. For those of us who don't usually think in terms of square kilometres, a few comparisons might help.

New York's Central Park is about 3.4 square kilometres, while Vancouver's Stanley Park is about 4 square kilometres. Central Park could fit on the Columbia Icefield 95 times, and Stanley Park would fit 80 times.

In some places, Athabasca Glacier measures 300 metres deep. That's as tall as the Eiffel Tower.

(1000 feet) deep. At this central point the glacier moves at a rate of 25 metres (82 feet) a year, gradually slowing down to 15 metres (49.5 feet) a year at its terminus. Although it is constantly advancing, however, it is still unable to keep pace with melt at its snout; the Athabasca has been retreating fairly consistently since it was discovered in 1898.

But even statistics can't fully represent the sheer mass of the Icefield. The glacier and its surroundings have an overwhelming effect on all but the most jaded traveller. The ice at the centre is as deep as the CN Tower in Toronto is tall. The glacier, its attendant icefield, and its surrounding high altitude environs cover an area the size of one of America's small states. The only man-made device that is able to rank with the Athabasca Glacier's slow but stupendous erosive capacity might be a large nuclear bomb. The icefield is so big that it creates its own winds and its own weather, and snow that fell more than two hundred years ago is just now melting at the glacier's snout.

There is a different sense of time here, only the timelessness of epochs, the incomprehensible vast passing of geological seasons, of mountains rising and falling, of the coming and going of the ages.

Ice Ages and the Columbia Icefield

As long as there has been ice in the polar regions of this continent, glaciers have likely existed in the Canadian Rockies. While high elevations and the resultant cold temperatures may have supported glaciers in the Columbia Icefield area for as long as three million years, ice movement here can also been linked to more widespread climatic coolings. These climatic changes resulted in major glacial advances throughout the Rockies.

What was probably the most extensive of all modern ice ages appears to have begun roughly 240,000 years ago. Back then, the Illinoisan or Great Glaciation covered most of the Earth's northern hemisphere, and shaped much of North America as we know it today. The Great Glaciation was a spectacular geological event that lasted 100,000 years. Though the glaciers of the Columbia Icefield

Alpenglow
The vibrant colours of alpenglow are a spectacular sight in the Rockies. Pink hues such as those seen above can appear in the sky just before the sun rises, or shortly after it sets.

would have grown dramatically during this continental cooling, the dynamics of the Icefield itself would not have changed much. Then, as now, warm, moist winds from the Pacific would deposit heavy snows along the divide between what is now British Columbia and Alberta. Snow would accumulate and compress into ice. Ice would begin to flow down valleys already created by ancient

Why Do Ice Ages Happen?

Little is known about why ice ages occur on Earth, but there are many theories on what might bring them on, including:
• changes in the orbit of the Earth
• variations in the heat output of the sun
• cosmic or galactic influences
• planetary collisions with cometary nuclei
• the sun-screening effect of airborne, iridium-laced ash
• the presence of visiting suns to our solar system
• variations in the thermal characteristics of the Earth's oceans
• the effects of volcanism on the Earth's atmosphere
• changes in the heat reflectivity of the Earth's surface

• changes in the latitudes of the continents
But perhaps the most likely local cause of glaciers is the mountains themselves. When moisture-laden winds are forced into the thinner atmosphere and colder climes of the high peaks, airborne moisture condenses into heavy snowfall, which fails to melt during the brief high altitude summers. The eternal snows linger in basins between the high peaks. Soon the snow is compressed by its own accumulating weight into glacial ice. The ice, by its very nature, then begins to flow downhill, in the direction of least resistance, into neighbouring valleys. It is clear that the Columbia Icefield owes its continued existence to the ring of high mountains that surrounds it.

rivers. The only probable difference now is in the dimensions of the major alpine glaciers. Long ago, they may have been hundreds of kilometres long and as much as two kilometres deep as they left the Rockies, where they joined the ice masses flowing southward from the direction of the pole.

Notable but lesser glacial advances took place in the Rockies 75,000 and 20,000 years ago. These advances did much to give these mountains the contours that make them so dramatic today. Another climatic cooling took place around 11,000 years ago and started what is called the Crowfoot Advance, a smaller glacial growth period still represented in the surface geology of the Columbia Icefield area. The last glacial advance to have taken place in the Canadian Rockies is so recent that early travellers were able to document its close. The Cavell Advance, often called the "Little Ice Age," likely began near the year 1200, roughly around the time that King Richard the Lion Heart was killed in France and the Fourth Crusade was making life miserable in Constantinople. At the peak of this advance, in about 1750, the Athabasca Glacier was two kilometres longer than it is now. Most of the other major glaciers that flow from the Columbia Icefield must have been much larger then, too.

Witnesses At the Edge of Time

The dynamics of glacial ice are difficult to study. We owe much of what is currently known about the Athabasca Glacier and the history of glacial advance and retreat in the Columbia Icefield

How Do Icefields Produce Their Own Weather?

It seems obvious that air is going to feel a bit chilly when it's sitting over a hulking block of ice, but an icefield can affect more than just the air directly above it.

If you visit the Columbia Icefield, you will undoubtedly notice a cold breeze coming from whatever glacier you're looking at. This cold breeze is caused by the same force of nature that causes skin to sag and rocks to roll downhill – gravity. Cold air is denser, and thus heavier, than warm air. As the air right over an icefield cools, gravity pulls it downhill, and you experience a glacial breeze.

During the day, glacial breezes usually combine with warmer regional winds. At night, however, regional winds often die down, leaving glacial wind free to collect in valley bottoms. These pockets of glacial air in the midst of warmer areas are sometimes called frost hollows, and can be several degrees colder than surrounding areas.

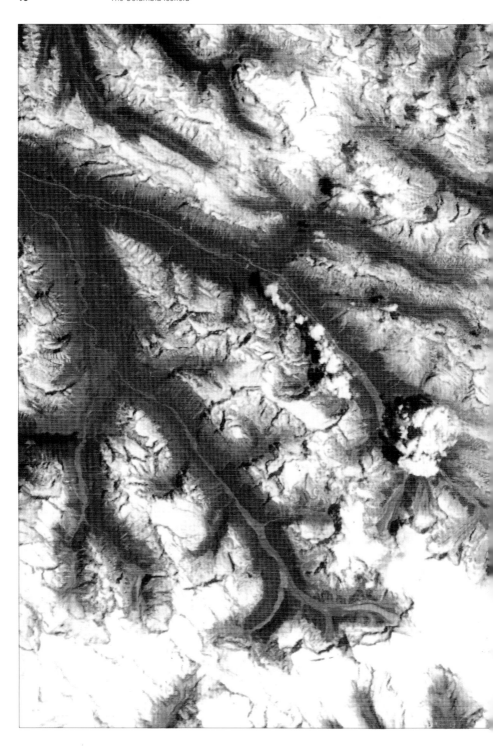

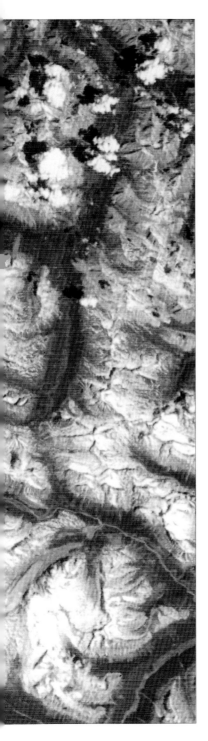

The Columbia Icefield from Space

This satellite photograph covers the southern portion of Jasper National Park (at top), the northern edge of Banff National Park (at bottom) and a small portion of Alberta outside the national parks (at right). The Columbia Icefield covers much of the bottom right quarter of the photograph. The Saskatchewan Glacier, with its prominent medial moraine, is quite obvious (lower right), as is the Athabasca Glacier (right centre) and the long, slender Columbia Glacier (top centre). Between these three prominent glaciers lies the Columbia Icefield, which feeds each them.

area, to a number of highly trained and very persistent researchers.

Though other scientists conducted earlier groundwork in the Icefield area, the work of Richard Kucera of the Department of Geological Services at the University of British Columbia has played a critical role in developing many of the concepts of glacial dynamics. In 1972, Kucera was in the Icefield area, carefully cataloguing glacial landform features. He also conducted time-lapse motion picture studies of the advance of the ice, which were later used in an Encyclopedia Britannica film called *Glacier On The Move*. The film offers an incredible glimpse into the most profound workings of nature.

Brian Luckman, member of the Univesity of Western Ontario's Department of Geography, has also made a significant contribution to the understanding of the Athabasca Glacier's dynamics. Dr. Luckman's research has focused on the larger patterns of climate and its impact on the glaciers and on communities of living things that surround them. At the snout of the Athabasca Glacier is a small ice cave. Here, Dr. Luckman discovered 8000-year-old wood fragments. Found in the summer of 1986, these fir and pine fragments seem to prove that, eight millennia ago, the glaciers of the Columbia Icefield were less extensive than they are today.

In Luckman's careful reconstruction of the climatic conditions during the recent "Little Ice

Icefalls and Seracs

(Above left) Where the irregularity over which a glacier must move is particularly large, the glacier will break apart to cross it. The breakup of a glacier as it flows over a steep cliff face below it creates icefalls. The Athabasca Glacier has three magnificent icefalls just below its headwall. The standing towers of ice that form the icefall are called seracs. As these ice towers are in constant motion, icefalls are very dangerous places to travel. After the glacier passes over these icefalls, its reconstitutes itself and continues to flow down-valley.

Glacial Streams

(Above right) At high altitudes, the virtually unfiltered summer sun melts winter snow from the surface of the glacier. Once that snow is gone, it begins to melt the ice itself. The resulting streams wander all over the glacier. As the cold, clear water hisses across the often knife-sharp ice, it gurgles and splashes like surf. Some of the greatest rivers on this continent have their origins in this melt.

Millwells

(Right) Rock debris that litters the glacier's surface has a greater capacity to absorb and retain the sun's warmth than the ice that surrounds it. Rocks often melt right into the surface of the glacier. Sometimes depressions carved into the ice in this way attract actively moving meltwater and become streams dropping from the surface of the glacier right into the heart of the ice. These holes, called millwells, or *moulins* in French, collect meltwater from the surface of the glacier and transport it into sub-glacial streams and rivers that course beneath the ice.

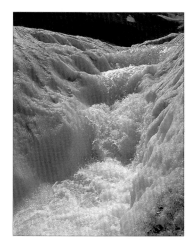

Crevasses

Under extreme pressure, glacial ice acts in the way a partial plastic might. A glacier, especially one like the Athabasca, is most plastic at its centre, where it is under the greatest pressure from its own weight and where the influence of drag, caused by the movement of the ice over the bedrock beneath it, is least. This plasticity allows the glacier to stretch. If a glacier stretches too much, however, it dissipates the internal pressures that give the ice its plasticity. The ice then breaks apart to form crevasses. Because of the nature of the ice, crevasses are rarely more fifty metres deep. (A crevasse is a fracture in ice. A crevice is a break in rock.)

The Headwaters of the Athabasca River
This aerial photograph features the northern section of the Columbia Icefield and the headwaters of the Athabasca River. The long, slender glacier at the centre of the image is the Columbia Glacier.

Age," he and his associates have conducted tree ring-aging studies in a number of mountain regions, including the Columbia Icefield. By studying annual growth ring development in trees that survived that local ice advance, Luckman and his associates are learning more about its climatic circumstances. In his pursuit of old trees, Luckman discovered a thousand-year-old larch snag at tree line near the Athabasca Glacier. Located 90 kilometres (55.8 miles) north of the range of the larch tree in the Rockies today, this ancient snag is further evidence that the Columbia Icefield area was warmer than before, and that its glaciers were smaller a thousand years ago.

The Future of Glaciers

It is very difficult to predict what will happen next in terms of glacial advance and retreat in the Rockies. Though much of what happens in terms of ice dynamics at the Columbia Icefield depends on local or micro-climatic conditions, even this great mass of ice is not immune to the larger influences of planetary climatic change.

Because of the way glacial ice is accumulated and endures, glaciers offer an excellent and reliable source of information about the Earth's past climates. Glacial cores from the ancient ice of the poles tell us that, before 1900, natural atmospheric pollutants overwhelmingly outnumbered human-generated pollutants. This is no longer true. Since 1950, for example, we have tripled the release of carbon dioxide into the atmosphere. In 1987 alone, carbon dioxide emissions from the burning of Brazilian rain forests exceeded those of all the fossil fuel use in all of North America and Oceania.

Over the last century we have altered the very nature and composition of the Earth's atmosphere. The carbon dioxide concentration in the atmosphere is now 25% higher than it was when

Photographing the Columbia Icefield

Most of today's sophisticated still cameras and video recorders have light metering systems that can read the unusually high reflectivity of glacier ice and compensate for it. If you do not have such equipment, however, you may have to know how to adjust your equipment so that it's not fooled by the extremely powerful light that is reflected from the glacier at mid-day, especially when the sun is shining brightly. Almost all cameras can be adjusted for this kind of full light.

In the box in which you purchase your film you should find instructions on how to expose the film in various kinds of light conditions. Most will tell you to over-expose an image taken of sand or bright snow. If the photograph you would like to take is composed largely of brightly lit ice, you may need to over-expose your photograph by as much as one to two f-stops relative to your light meter reading. The best way to ensure that you are not disappointed with your close pictures of the glacier is to try two or three different exposures. Start with the normal exposure indicated by your light metering reading and then over-expose by one f-stop, and then by two f-stops. Depending on the angle of the light and the brightness of the ice, you will be guaranteed of at least one perfect exposure.

The Glacial Surface

At first glance, the surface of the Athabasca Glacier may appear smooth and regular. But late-lying snow often disguises the true nature of the glacier's surface. Glacial ice, especially when it has been exposed to surface melt, is very rough and highly abrasive. In summer, meltwater streams wander all over the ice. Where the glacier flows over irregularities in the bedrock beneath it, the ice is apt to break into fractures called crevasses.

the first explorers visited the Columbia Icefield not even a hundred years ago. Industrially-caused acid rain has increased spectacularly, as have the concentrations of methane and chloroflourocarbons. New holes in the highly protective ozone layer further threaten our atmosphere's capacity to sustain life as we know it on Earth. Other human activities have already accelerated the rate of species extinctions to a thousand times its geological norm, and large scale changes in global weather could further accelerate the disappearance of life from this planet.

Global Warming

Human activity is now the single greatest factor influencing the evolving climate of the Earth. One cumulative effect of all of these atmospheric changes appears to be a "greenhouse effect" that is causing the planet's climate to warm. The ultimate impact of global warming will depend on its extent and duration, but it's clear that such warming could have a substantial impact on the Earth's inter-connected ecosystems and on patterns of human settlement. It's simply not known what the ultimate effects of these atmospheric changes will be on the world's glaciers.

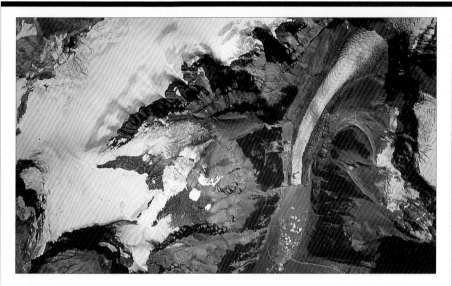

Great Moments in Columbia Icefield History

Scientists can determine periods of significant glacial advancement and decline by reading the geological record of rock and ice. In the Canadian Rockies, this record goes back at least as far as the period between 238,000 and 126,000 BC, when icefields like the Columbia Icefield were experiencing major growth.

During the Great Glaciation, or Illinoisan, period from approximately 238,000 to 126,000 BC, Neanderthals had already begun the ritual of burying their dead, one of the first signs of emotion or thought placing humans apart from other animals. Toward the end of the period of advancement called the Early Wisconsinan, from about 73,000 to about 62,000 BC, Homo sapiens started to appear on Earth.

The next major advance, the Late Wisconsinan, occurred between approximately 18,000 and 9000 BC. In this 9000 year period, humans in the Middle East domesticated the dog. They also discovered the advantages of herding animals over just hunting them. The end of this glacial period saw the beginning of the Sahara Desert and the disappearance of intercontinental land bridges.

During the time of the Crowfoot advance, which began after 9000 BC and ended before 7000 BC, humans learned to farm, the Biblical city of Jericho was founded, and the development of irrigation led to the rise of city-states along the Tigris, Euphrates, and the Nile Rivers.

The Little Ice Age, or Cavell, advance lasted from about 1200 AD to 1900 AD. During this period, the Athabasca Glacier peaked, Europe was embroiled in the War of Spanish Succession, Daniel Defoe was writing *Robinson Crusoe,* and J.S. Bach was composing some of his greatest organ music. The Athabasca Glacier then receded until about 1800, when it started to advance again, almost regaining its 1715 size in 1840 – about the same time photography was being invented.

Visiting the Columbia Icefield

Early Snowcoaches
A precursor of the present-day Snowcoaches, which wheel visitors right up to the impressive Athabasca Glacier via the Icefield Parkway.

To be at the toe of a glacier during peak melt is to understand the full extent to which glaciation has molded this continent. If you wait long enough, and slow your own pace down at least to the extent to which any animate creature can slow in sympathy with the ages, then the pinging and groaning of the advancing ice can, at last, be heard. This is not just the sound the glacier makes as it grumbles toward the valley floor, this is the sound that ice has made for more than three million years in these mountains as glaciers of the Columbia Icefield have shaped and reshaped the surface of the land.

But there is more than just the echo of the glacier scraping forward across the naked bedrock over which it has likely flowed for millions of years. There is the sound of water, too. On any given summer day, over the entire surface of the glacier, you will see that water has thinned the ice. That water accumulates into surficial streams. Rivers are then created beneath the ice by way of crevasses and millwells that channel the waters to the bottom of the glacier. From here, the waters are ultimately expelled at the glaciers' toe.

It is at the snout of the glacier that rivers form. In the case of the Columbia Icefield Glaciers, these rivers are substantial. Few visitors to the toe of the Athabasca Glacier fully appreciate what they are witnessing here; right before their eyes is the watershed of a continent. Travelling to this glacier is like travelling back in time, to the middle of the last glaciation. To see the glacier as it is today, to feel the wind that is drawn down-valley by the cold on the icefield above, to hear the sound of the melting water, is to know what most of the northern half of this continent must have been like during the time of the last great ice age. But this privileged access to the primacy of the past is not without hazard. The toe of the Athabasca Glacier is a dynamic and dangerous place.

Life and Death on the Ice

Accidents are not uncommon on the Athabasca Glacier. There is a story told by those who know the area of a family that drove to the parking lot near the toe of the Athabasca Glacier one summer. Perhaps because of their awe, and maybe because so many others were anxious to touch the melting ice at the toe of the glacier, this family went far past the Canadian Parks Service signs warn-

Safety Tips for Visitors

The Icefield is not an inherently dangerous place, but it can be hazardous, especially for the inexperienced. Fortunately, a modicum of caution and good sense significantly reduces your chances of being hurt while visiting.

1. The most important, and most often ignored, safety tip is to stay off the ice. Although you may see many people walking around on the ice, especially at Athabasca Glacier, anyone who is not an expert at glacial hiking should not actually be on the ice without an expert guide. From April through early July, most of the Icefield is covered with snow, which forms insubstantial bridges over crevasses. These bridges can be very difficult to detect, and they collapse easily under pressure. When the snow is gone the crevasses are easier to avoid, but the ice is extremely slippery. Injuries caused by simply losing one's footing and falling on the hard ice are quite common and can be serious.
2. Stay on the trails. According to the Jasper Warden Office, walking on the gravel slopes around the Icefield can feel like trying to walk on marbles coating a really hard surface. Keep in mind that the whole Icefield area is constantly in flux and nothing is in an absolutely permanent position. Even very large boulders can be loose. Straying from the trails is not advised.
3. Dress warmly and wear sturdy shoes with good traction. It's always a little colder around an icefield than in surrounding areas, and it's no fun to shiver and turn blue when you're trying to take a picture. If you do walk on the ice with an expert guide, sturdy shoes are essential.
4. Keep children in sight and under control. The Icefield is not like a backyard or a playground.
5. Don't throw rocks. Rocks bounce around a lot more on ice, and tend to displace other rocks.

Walking and Hiking Guides
(Opposite) Expert guides conduct walking and hiking tours for those who wish to have a challenging experience in the Columbia Icefield area.

ing visitors of the enduring hazards of glacial ice. Seeing no apparent hazard in the cracks of the glacier's terminus, they went right up onto the ice. Seeing others walking in the area above them, the father and the son continued on. Suddenly, the son disappeared. Quickly making his way to where the boy had vanished, the father found his son wedged, out of reach a few metres below, between the jaws of a blue crevasse.

Try as he might, he could not reach the boy. He called a nearby mountain guide, who was leading a group of sightseers on the glacier. The guide quickly tried to reach the boy. Though the guide was able to attach a rope to the child, he was unable to gain the leverage necessary to pull the boy out of the crevasse. A national park rescue team was summoned.

It was an hour and a half before the boy was finally pulled from the crevasse. By that time, the ice had cooled the core temperature of the boy's body – he had already become hypothermic, and soon died.

A legal battle ensued, during which the boy's bereaved family sought to lay the responsibility for the death of their son on the Parks Service. But the Parks Service did not kill the boy. It was no

Journey Beneath the Ice: A Survivor's Tale

A couple of decades before this book was written, I was a young naturalist with the Canadian Parks Service. During my first season working in Banff National Park, I took one of my summer weekends off to backpack over the Saskatchewan Glacier into the famous Castleguard Meadows, the birthplace of the longest cave system in Canada. This half-million-year-old cave system extends from unknown sources beneath the Columbia Icefield to an outlet at the edge of the meadows beneath Mt. Castleguard, one of the most remote places in all of Banff National Park.

The first leg of this journey was up the gradual slope of the Saskatchewan Glacier and then across its bergschrund and into the high alps of the meadows. I had no previous experience on glacial ice. While my well-seasoned companions scampered up the smooth grade of what they thought was an easy glacier, I was left on my own to deal with the terror of my first glacier ascent. It was a very long journey for someone who had never backpacked before.

As I had only two days off, I had to start for home the next day. An experienced hiker was sent down the glacier in my company but, not knowing how uncertain I was, he left me far behind. Near the toe of the glacier, with my partner almost out of sight in front of me, I plunged into a deep surficial stream and was carried away by the force of the rushing water. The water poured into a shallow crevasse that took me under the glacier.

As the stream carried me under the ice, it grew dark. There was a space of only a few centimetres between the top of the water and

the roof of the ice. I remember trying to take my backpack off, but the water fought against that idea and, besides, other events soon distracted me. The roof of the ice had begun to glow. At first it glowed a faint green, and then an icy blue. Just as I became captivated by the colours, I burst out into bright sunlight at the toe of the glacier. I was on the wrong side of the newly formed river, but alive.

Miraculously, the quick flush through the ice had allowed me to catch up with my companion, who had turned around just in time to see me fall into the ice. With his help I was able to cross the main stream of the river to where the trail made its gradual way to the highway five kilometres away. To my great distress, my experienced companion then declared that the ordeal had been too much for him, and that he had elected to stay with a survey crew on the outwash of the glacier. While he dealt with his shock, I considered the fact that I was expected to work the next day. One of the surveyors journeyed with me through the gorge of the Saskatchewan to within sight of the road.

Alone at the edge the highway, no motorist would pick me up, even when I stood in the middle of the road. At nightfall, two women from Quebec, who spoke no English at all, stopped and saved me. It took me months to recover from the shock and the cold. A few years later I was one of the first park naturalists to guide visitors on walks on the surface of the Athabasca Glacier. To each visitor who joined me on the rough surface of the ice, I tried to offer more than just statistical facts. The geology of the Columbia Icefield had penetrated me, and touched my very soul.

Snow Machines

(Photos top to bottom)

1. The Bombardier

Post-war experiments with tracked vehicles proved these machines were safe for travel on the Athabasca Glacier. The evolution of this technology is an interesting study. This snowmobile, manufactured by Bombardier in Canada, carried six to eight passengers and was used on the Athabasca Glacier until the early 1980's.

2. A Bus On Tracks

As group tour travel became more popular, larger vehicles were required. The first experiments with larger snowcoaches saw bus bodies attached to track mechanisms developed for arctic oil exploration.

3. Special Designs

As the Athabasca Glacier tour became more and more popular, it was seen that a custom-made solution to the problem was needed. This tracked vehicle was the intermediate step between the snowmobile and the modern snowcoach.

4. The Snowcoach

The realization that tracked vehicles cause unacceptable disturbance to the surfaces over which they travel led to the evolution of all-terrain vehicles with large, low-pressure tires. This technology was tested in the arctic and found to be far more efficient and environmentally acceptable than earlier tracked designs. The Foremost Terra Bus, or Snowcoach (also Snocoach) was developed jointly by Brewster Transportation and Tours and Canadian Foremost in Calgary, Alberta. Though very expensive, these 56 passenger machines are safe, comfortable and highly reliable.

one's fault. It was the callousness of eternity that killed the child, the blind coldness of the indifferent ice. As this story shows, all visitors to the toe of a glacier should bring with them a profound respect for the ice, and be aware of its hazards at all times.

Snowcoach Tours and Guided Hikes

Beyond the trained mountaineer's knowledgeable approach to the glacier, there are two ways in which the average visitor can actually exprience the ice. The first is the snowcoach tour, the second a conducted walk. Each of these is a valuable vehicle for safely visiting and understanding the ice.

Prior to the completion of the Icefield Parkway, few people had ever seen the Athabasca Glacier. Fewer yet had seen the icefield that had formed it. After the parkway was officially opened in 1940, the idea of mechanized public transport on the readily accessible Athabasca inevitably began to take shape. In 1946, an army half-track vehicle made its appearance on the ice. In 1948, a man named Alex Watt began offering tours on the glacier using half-tracks, a project that succeeded well enough to draw the attention of the Parks Service, who later granted a concession licence for the operation. Much to Watt's disappointment, a Jasper entrepreneur named Bill Ruddy was awarded the official government concession for the Icefield in 1952. It was Ruddy who introduced the famous six passenger Bombardier

Riding the Snowcoach

Specifically designed for Brewster Transportation and Tours for use on the Athabasca Glacier, the Foremost Snowcoach· is manufactured by Canadian Foremost in Calgary, Alberta. Equipped with large, low-pressure Terra tires, the six wheel drive snowcoach can carry 56 passengers with a minimum impact on the surface of the ice.

Dimensions

Height	3.86 m (12 ft 8 in)
Length	13.0 m (42 ft 8 in)
Width	3.61 m (11 ft 10 in)
Weight	19,500 kg (43,000 lbs)

Powertrain

Engine	DDA 6V71, 210 HP (155 kW) @ 2100 RPM
Transmission	Clark 2800 Series Powershift
Suspension	Leaf spring – front Walking beam – rear
Axles	Clark Drive Steer Planetary – front Clark Planetary – rear
Tires	Goodyear Terra Tires 66 X 43.00 X 25
Brakes	Air over hydraulic front drums; air rear drums

Performance

Speed	Low range	18.4 km/hr (11.4 mph)
	High range	42 km/hr (26 mph)
Turning radius		16.8 m (55 f)

snowmobile to the surface of the glacier. Ten years later, the Banff-Jasper highway was paved, changing the complexion of visitation to the icefield for good. In 1969, Bill Ruddy sold Snowmobile Tours Ltd. to Brewster Transportation and Tours, the legendary Banff sightseeing company.

In purchasing the Athabasca Glacier snowmobile operation, Brewster inherited the problem of developing more reliable and more easily maintainable technology for mass transport onto the ice. Brewster experimented with a range of new technologies, including placing highway coach bodies on complicated track mechanisms. The company solved the problem in 1981 when they contracted Foremost, a Calgary company that made vehicles for arctic oil exploration, to develop a special vehicle for

10 Popular Questions About the Columbia Icefield

1. Is it the Columbia Icefield or the Columbia Icefields?

Because the Columbia Icefield is a basin out of which glaciers flow, it is considered a singular geological feature. Out of this major single icefield, however, flow many glaciers.

2. Why is the ice blue?

Glacial ice is composed of compressed snow. At the centre of every snowflake is a nucleus of dust. Because of the size of the dust particles and the nature of the crystalline structures in which they are trapped, the ice reflects the shorter blue and green wavelengths of light.

3. Can you drink the water?

The water created by the direct melting of glacier ice is almost as pure as distilled water. Clear glacier water is as fresh and clean as any water on Earth. When glacial melt picks up rock flour and other debris, its colour turns from clear to grey or brown. Though the water is still fresh, these sediments can bother sensitive stomachs if the water is consumed in large quantities.

4. Is the glacier advancing or retreating?

By its very definition, glacial ice is in constant motion. A glacier only retreats when it melts away faster than it advances. Though the Athabasca Glacier is constantly advancing, it melts back faster in the summer than it can advance in the winter. Thus the Athabasca is said to be retreating.

5. Are all the glaciers in the Rockies retreating?

No. Though most glaciers appear to be retreating, glaciers at the higher altitudes do not appear to be impacted by the general warming of the climate in the Icefield area over the past century. Few glaciers, however, seem to be in any form of advance. Though the Columbia Glacier was seen to be advancing slightly between 1966 and 1977, that activity seems to have halted.

6. Does anything grow or live on the ice?

Yes. In spring and summer, late lying bands of snow turn pink. Hikers walking through this snow often leave red footprints. This coloration is caused by a blue-green algae of the species Clamydomonis nivalis. The microscopic cells of travel on the Athabasca Glacier. Though very expensive to develop and build, the resulting fifty-six passenger snowcoach easily handles the steep hill down the knife-edged lateral moraine onto the ice. It also gives everyone a once-in-a-lifetime opportunity to walk off the coach, onto the ice, right at the centre of the Athabasca Glacier.

Given good weather, it is possible to take a guided walk on the surface of the glacier. Initially undertaken by the Canadian Parks Service as part of the Jasper National Park Interpretive Program, guided walks on the glacier are now offered by the Athabasca Icewalk company. Led by highly trained mountaineers well-versed in glaciology, these glacier walks offer safe and interesting experiences that start at the toe of the Athabasca Glacier.

---

this algae are encased in a red gelatinous sheath which is capable of withstanding cold temperatures, and which may also protect the algae from the fierce radiation that falls on snow at high altitudes. Some say this snow, if eaten, has the faint taste of watermelon.

Many insects can also be found living out various stages of their lives in the snow. Iceworms also exist in the Rockies, but they are not common here.

7. What does Athabasca mean?
This word comes from the native Cree language and means "place where the reeds grow." This name refers to the Athabasca River in the area of Lake Athabasca in northeastern Alberta. The Athabasca glacier melts to create the Sunwapta River, which joins the Athabasca not far from the Columbia Icefield.

8. What is this area like in winter?
The Columbia Icefield area receives a great deal more snow than any of the areas immediately around it. An average snow year would see ten metres of snow fall on the Athabasca Glacier. It is also very windy and cold in this part of Jasper National Park. Temperatures of -40 to -50° Celsius (-104 to -122° Fahrenheit) are not uncommon.

9. What kinds of animals survive the bitter winters?
Most of the animal residents of the mountainous areas of Canada migrate away from or hibernate through the long, cold winters at the edge of the permanent ice. The few that don't, like bighorn sheep and mountain goats, have evolved sophisticated physiological adaptations to cold and wind. They are among the toughest creatures on Earth.

10. Can you go out on to the ice?
Yes and no. Any glacier can be a very dangerous place if you don't clearly understand or can't identify the hazards. Many people have lost their lives by going on to the glacier without the right equipment and knowledge. If you are a mountaineer with considerable previous experience on glacial ice, you can register for climbs or explorations on the glacier and on the icefield above. If you are not experienced, it is suggested that you go out on to the ice under the care of a trained mountain guide or that you take a snowcoach tour. Information about glacier tours is available within sight of the ice.

A Visual Tour of the Columbia Icefield

Waves of Peaks (Above) The Canadian Rockies block warm, moisture-laden winds that blow from the Pacific. As these winds rise to pass over the peaks, the air is cooled and moisture falls as rain or snow. The Columbia Icefield is unique in that there are no other major mountain ranges between it and the Pacific. A gap in the interior ranges of British Columbia makes this part of the Great Divide the first major obstacle to winds from the west. This accident of geography has helped create one of the most remarkable natural features in Canada.

The Great Divide (Opposite top) The Great Divide is the spine of the Canadian Rockies, and the highest range of peaks blocking winds from the Pacific. The Great Divide separates those rivers that flow into the Pacific from those that flow into the Atlantic. The Great Divide also separates Alberta from British Columbia. It is on the Great Divide that the greatest snowfall occurs in the Rockies.

Mt. Bryce (Opposite bottom) Eleven of the twenty-two highest peaks in the Canadian Rockies are found in, or in close proximity to, the Columbia Icefield. Though not as high as the Rockies in the United States, these mountains are found at a higher latitude. Treeline is lower and more of the mountain is exposed to the cold. For this reason most of the Great Divide is bare rock, scree and ice. The configuration of the peaks in the Columbia Icefield area creates a high basin in which snow accumulates year after year. It is in this basin that the Columbia Icefield rests.

Permanent Winter

(Top) Short summers combine with an average snowfall of nearly 10 metres (30 feet) a year to create the Columbia Icefield. High altitude ensures that much of the snow that falls in winter doesn't melt in summer. The snow that falls annually is gradually compressed by its own weight into glacial ice. It is this flowing ice that forms the nine major glaciers that flow from the Columbia Icefield. The distant mountain is Mt. Bryce.

The White, Eternal Snows

(Above left) By any standard, the Columbia Icefield is an impressive geographical feature. It is a high basin of accumulated snow and ice that straddles 325 square kilometres (roughly 125 square miles), of the Great Divide. An icefield is an area of annual snow accumulation. A glacier is compressed ice that begins to flow out of an icefield and into a neighbouring valley. This photograph shows the Columbia Icefield basin and a glacier as it flows downhill out of it. Mt. Columbia is in the background.

Mt. Columbia

(Above right) Mt. Columbia, on the northern edge of the Columbia Icefield, is the highest mountain in Alberta. Its south slopes are almost completely buried in glacial ice.

The Saskatchewan Glacier

The longest outflow from the Columbia Icefield is the Saskatchewan Glacier. It is nearly seven kilometres (over four miles) long. The Saskatchewan Glacier is the headwaters of the North Saskatchewan River. Its surface is marked by a medial moraine. Here two glaciers merge to form a single common lateral moraine that looks like a highway dividing the two glaciers. The moraine, however, is composed of broken rock and is difficult enough to walk on, let alone drive.

The Columbia Glacier

(Above and opposite) The Columbia Glacier is one of the most spectacular glaciers that flows from the Columbia Icefield. This glacier, which is the headwaters of the Athabasca River, leaves the icefield through a narrow gap and falls steeply into the Athabasca River Valley adjacent to the north face of Mt.

Columbia. The Columbia Glacier is readily identifiable by the pressure ridges, or ogives, that pattern the surface of the ice. Ogives (pronounced oh-guy-ves) are bands or undulations of darker and lighter ice that appear in periodic patterns on the surface of a glacier. These patterns mark the summer and winter advance of the ice.

Mt. Andromeda

(Top) Although there were earlier ice ages, it is thought that the Columbia Icefield has existed much as it is today for three million years. Mt. Andromeda, which towers over the Athabasca Glacier, has been given its dramatic shape by the grinding of glacial ice over long periods of time. Mt. Andromeda is surrounded by the ice of the Columbia Icefield and is also draped by its own hanging glaciers.

Mt. Alberta

(Bottom) Though not technically within the Columbia Icefield proper, one of the great peaks of the icefield area is Mt. Alberta. This mountain was the last major peak in the Canadian Rockies to be climbed. It was considered unclimbable until it was finally scaled by a Japanese expedition in 1925.

Mt. Athabasca

(Top) Mt. Athabasca is the mountain most often associated with the Athabasca Glacier. It has been carved by the action of three interacting glaciers into the shape of a "matterhorn" peak. This mountain is probably the most popular mountain in the Canadian Rockies for novice climbers. It was from the summit of this mountain that the Columbia Icefield was discovered.

Mt. Athabasca and Little Athabasca

(Bottom) This lower southern outlier of Mt. Athabasca is often referred to as Little Athabasca. It, too, has been shaped by glacial ice into a matterhorn. This is the view of Mt. Athabasca as seen from Parker Ridge. A trail past timberline on Parker Ridge offers easy access to fine vistas of the Columbia Icefield's southern outflows.

Athabasca Storm-rise

(Middle) The Columbia Icefield is large enough to create its own weather. Warm, westerly winds rise and are cooled by the great expanse of ice. These winds pour over the divide and down the Athabasca Glacier into the Sunwapta Valley. Snow storms sometimes occur even during the summer months. Such storms can be spectacular in the effects they produce.

Parker Ridge

(Top) It is from Parker Ridge that the impressive Saskatchewan Glacier can be viewed. The trail up Parker Ridge is short and very spectacular. The alpine regions through which the trail wanders, however, are very fragile, so please stay on the trail. The alpine tundra zone on Parker Ridge is excellent habitat for mountain goats and bighorn sheep. Occasionally grizzly bears can also be spotted from the viewpoint at the crest of the ridge.

Dome Glacier

(Bottom) Dome Glacier flows into the Sunwapta Valley near the Athabasca Glacier. The mountain from which it flows, Snow Dome, is the watershed apex of western North America. The summit of this mountain is a triple continental divide. Meltwater off this mountain flows into the Atlantic Ocean from one side, the Pacific from another, and the Arctic Ocean from still another of its slopes.

Tangle Falls

(Opposite) Tangle Falls is one of the many spectacular waterfalls near the Columbia Icefield.

The Athabasca Glacier

(Top) The Athabasca Glacier has been receding in size almost continuously since the beginning of the twentieth century. In the early 1920s the toe of the glacier was very near the current location of the Columbia Icefield Chalet. Though global warming may affect glacial dynamics, it is difficult to predict the future activity of the glacier.

Glacial Silts

(Right) Highly compressed glacial ice contains fine particles of dust. When glacial ice melts, this dust is released and mixes with water and powdered "rock flour" created by the grinding of the ice over the bedrock beneath it. These fine powders make for the creation of mud that has the quality of quick sand. Though seldom deep enough to threaten a human, a slip into this mud can ruin a visit to the glacier's snout – a good reason to stay on the trail.

Snowcoach Tours

(Left) Many visitors to the Athabasca Glacier pay to ride a specially designed snowcoach onto the surface of the glacier. This is a very safe and comfortable way for visitors of all ages and levels of fitness to experience the living surface of the ice. These interesting, fifty-six passenger machines travel along an ice road to a turn-around point just below the last major icefall on the Athabasca Glacier.

Glacier Walks

(Top) Unless you are an experienced mountaineer, it is not safe to walk on the surface of the glacier alone. For this reason guided walks are offered along safe routes on the Athabasca Glacier during the summer season. Weather depending, reasonably fit hikers can see many of the most spectacular glacier features at close range. Tours are limited in size and are offered by trained and knowledgeable mountain guides.

The Icefalls

There are three major icefalls on the Athabasca Glacier, each of which is created by a cliff-face beneath it. As the semi-plastic glacier flows over each of these cliffs, the ice stretches and breaks into deep crevasses. Though snowcoach passengers are not permitted to explore these dangerous rents in the glacier, the grumbling and groaning of the tortured ice can often be heard from the end of the snowcoach road.

Hanging Glaciers

(Above) Big alpine valley glaciers, like the Athabasca, flow out of the main mass of the Columbia Icefield. Though these classic glacial forms attract most of the visitors' attention, they are not the only glacial forms that in the icefield area. The mountains of this region are very often large enough and high enough to perpetuate their own glacial circumstances. This hanging glacier (left) on Mt. Athabasca, for instance, was likely connected at one time to the major flow of the Athabasca Glacier below it.

Late-Lying Snow
In glacial environments, growing seasons are very short. It is not uncommon for winter snow to remain well into the summer. This photograph was taken from the parking area at the Columbia Icefield Chalet in mid-July of 1991. Employees and regular visitors to this area often argue that there are only three seasons in this high, barren part of Jasper National Park: the very long winter, the brief, fierce spring, and autumn. Despite the cold, wildflowers bloom riotously in the late spring.

Autumn, Then the Long Winter

The Athabasca Glacier gets its name from the Athabasca River, a major western Canadian watercourse which originates in the Columbia Icefield. Athabasca is a Cree Native word that means "place where the reeds grow." While reeds do grow for half the year along much of the fabled Athabasca River, the growing season in the upper reaches of the Athabasca drainage is brief.

Life in the Columbia Icefield

Mountain Heather
The highest regions of the Rockies are, in effect, a southern extension of the conditions found in the Canadian arctic. Rarely are there fewer than seven consecutive frost-free nights a year in the high alpine regions of the Columbia Icefield area. For this reason, plants have to huddle together to protect themselves from the cold.

The peaks that surround the Columbia Icefield are the ancient ruins of an ocean. Powerful, conflicting forces in the earth elevated the sea floor and exposed the sediments to the gnawing wind. It is not known exactly what turned the great stone floor on edge to form mountains, but the forces were immense. Only the shifting of continents, it seems, could have so altered the crust of the Earth.

The heights reached by the rising, crumbling sea floor became a foremost obstacle to life. In effect, the highest regions of these mountains became a southward extension of the Arctic. The tundras of the north and of the high Rockies are made up of similar kinds of plants growing under the same rising and falling curtain of cold.

In the valleys, trees were quick to grow and forests spread up the sprawling shoulders of mountains. But at a certain point even the heartiest trees are hard pressed to survive the tearing winds and the long winters. The highest survivor is the alpine fir, which gathers into clusters around a "mother tree" and grows branches right to the ground to resist upward drafts of wind. It will grow along stone walls and on slopes, where it can take advantage of the contours of the land to reduce abrasion during storms. But at a certain altitude, all its devices fail, and blizzards blow the trees apart.

Treeline

The upper limit of the forest, known as treeline, is an almost inflexible barrier to life. The plants of the forest struggle up the shoulders of mountains but are finally repelled by the cold, the short summers and the naked faces of the peaks. But life does extend above the trees. Through gradual adaptation, living things have come to thrive at the bitter upward limits of life on Earth.

Outwardly, the plants that survive above treeline may seem little different from the forms of the lower valleys. They are smaller, perhaps, dwarfed by the huge peaks and the acid cold of frost that falls regularly even in summer. But the capacity to grow close to the ground is only one trick that plants use – and frost is only one murderous agent in the passive resistance of indifferent mountains. Late-lying snow often covers the ground until early July. Plants flower and then go to seed in the declining light of a greatly shortened summer. But alpine plants have found chinks in the great armour of the tundra cold. They ripen their seeds in the warm sun of late summer. The

seeds store food over the winter and then, in spring, they burst, using the power of last year's sun to push through the cold crust of the Earth and through the snow above it. The cycle is then repeated.

But even in the warmest summers, the growing season may be as short as 45 days. In cold summers, plants may not generate enough energy to produce a well-formed seed. So they wait. Some plants may take 20 years to produce a single flower. Of the more than 300 plants that grow in the alpine regions of the Rockies, only a few are annuals – the rest grow perennially so they don't have to replace everything each short summer.

To survive in the shadow of the ice, plants must grow slowly.

Alpine Buttercup
Alpine plants have found chinks in the great armour of the tundra cold. They ripen their seeds in the warm sun of late summer. The seeds store food over the winter. Then, in spring, they burst using the power of last year's sun to push through the cold crust of the earth. With this strategy, the alpine buttercup can blossom right through the spring snow.

The Western Anemone

(Top and middle) One of the most common flowers of the high alpine that surrounds the Columbia Icefield is the western anemone. It has hairs on its stalks to insulate the plant from the cold. It sets its seeds by broadcasting them on the wind after the seedhead has matured. Of the three hundred or so plants that grow in the tundra of the Canadian Rockies, only a few are annuals. All the rest, like the anemone, grow perennially so they don't have to replace all their foliage each brief summer.

Alpine Forget-Me-Not

(Bottom) Many alpine flowers are simply dwarfed versions of wildflowers that grow at lower altitudes. One of the most striking of the tundra wildflowers is the alpine forget-me-not. This small but stunning blossom is of the same unforgettable blue as the glacial lakes that dot the Great Divide.

A mat of leaves a hand's breadth across may be half a century old. A wind-torn fir may have been a seedling before Europeans ever saw these mountains.

Plants are married to the soils they grow in. The kind of soil found in any given area near the Icefield can usually be determined by the kinds of plants that grow on it. Communities of mountain avens are found on wild-dried slopes where soils are rich in calcium. Heather and everlasting communities intersperse on well-drained and stable surfaces, while communities of saxifrage grow where soil is covered by late snowbeds on north-facing slopes. Saxifrage is a latin term that describes the plant's ability to break rock down into soil.

The plants of the alpine are seldom solitary. By clustering together even into small communities, plants have a better chance to alter slightly the climate immediately around them. Huddled together, the plants of the tundra may stay warm enough to produce flowers.

After the snow melts, winds blow endlessly down the glaciers and across the high tundra. In the dry months of July and August, the winds rob the soil of moisture from spring melt. At the height of summer, a drought occurs. But the plants of the tundra resist the winds of the high desert – they trap the dew with hairs on their stalks, and from their succulent leaves the wind can draw no moisture.

As the sun burns down through the clear air of the high meadows, unfiltered cosmic rays and ultraviolet light pierce the tissues of every living thing. Sunlight is a second kind of wind. It is a solar wind, a great storm through which the Earth spins as it passes through the sun's glow. At the peaks, the colours of alpine flow-

The Ten Highest Icefield Peaks

Mt. Columbia: At 3747 metres, Mt. Columbia is the highest mountain in Alberta. It was first climbed in July of 1902 by James Outram and Christian Kaufmann by way of the southeast face. It was named by Norman Collie in 1898 for the Columbia River.

North Twin: At 3730 metres, the remote North Twin remained unclimbed until July of 1923, when it was ascended by W. S. Ladd, James Monroe Thorington and the famous guide Conrad Kain. This mountain was named for its similarity in shape to its twin, listed below.

South Twin: The 3580 metre South Twin was first climbed in July of 1924 by F. V. Field, W. O. Field, L. Harris, J. Biner and the Swiss guide Edward Feuz Jr.

Mt. Bryce: This magnificent 3507 metre peak was first climbed by James Outram and his guide Christian Kaufmann in July of 1902. It was not climbed again for nearly sixty years. The mountain was named by Norman Collie for James Bryce, the president of the Alpine Club in England.

Mt. Kitchener: This 3505 metre glacier-clad mountain was first ascended in July of 1927 by J. Ostheimer with his guide Hans Fuhrer. The mountain was named by the Geographical Board of Canada in honour of Horatio Herbert, First Earl of Kitchener of Khartoum and of Broome.

Mt. Athabasca: This 3491 metre peak was first climbed in August of 1898 by Norman Collie and Herman Woolley, the discovers of the Columbia Icefield. They named the peak for the Athabasca River.

Mt. King Edward: This 3474 metre peak was first climbed in August of 1924 by J. W. A. Hickson, Howard Palmer and their guide Conrad Kain. Howard Palmer and one of his companions named the mountain for King Edward VII.

Snow Dome: The summit of this hydrological apex is 3456 metres. It, too, was first ascended by Norman Collie, Herman Woolley and Hugh Stutfield on their first expedition to the Columbia Icefield in August of 1898. The trio named the mountain for its shape.

Stutfield Peak: The 3450 metre summit of Stutfield Peak was not reached by climbers until June of 1927, when A. J. Ostheimer and Hans Fuhrer made the ascent in a 36 hour mountaineering epic which saw the ambitious pair climb the North Twin, Stutfield Peak, Mt. Kitchener and Snow Dome without so much as a return to camp. The pair were on their way to Mt. Columbia when bad weather forced an end to their now famous enduro. Norman Collie had named the mountain for his friend and mountaineering companion Hugh Stutfield in the summer of 1898.

Mt. Andromeda: This beautiful 3444 metre summit was named by pioneer ski mountaineer Rex Gibson for the wife of Perseus in Greek mythology. It was first climbed in July of 1930 by W. R. Hainsworth, J. F. Lehman, M. M. Strumia and N. D. Waffl.

Spotted Saxifrage

(Top) The kind of soil found in any given area near the Icefield can usually be determined by the kinds of plants that grow on it. Heather and Everlasting communities intersperse on well-drained and stable surfaces while communities of saxifrage grow where soil is covered by late snowbeds on north-facing slopes.

Indian Paint Brush

(Bottom) After the snow melts, winds blow endlessly down the glaciers and across the high tundra. In the dry months of July and August, the winds rob the soil of moisture from spring melt. At the height of summer, drought occurs. But the plants of the tundra resist the winds of the high desert. With hairs on their stalks they trap the dew, and from their succulent leaves the wind can draw little moisture.

ers are vibrant and revealing. There are a thousand shades of blue and red, each telling of different acids and bases in the flowers' stalks.

Animals and the Ice

Compared to the lower valleys of Jasper National Park, there aren't many animals living in the Columbia Icefield area – the altitude is too high and the summers too short to support the abundant plant life needed to satisfy the appetites of elk herds, for instance. But keep an eye out for signs of life, because there are lots of creatures who call the Columbia Icefield home, or who at least pass through the area now and then.

The most commonly seen of the larger animals is the bighorn sheep, which can often be seen in small bands, sometimes including young, near Tangle Falls. Tangle Falls is located at the top of the big hill climbed by the parkway on its way to Jasper from the Athabasca Glacier. On rare occasions, a moose may follow the tasty willow bushes up valley past timberline. Black bears and even grizzlies are also visitors to the Columbia Icefield area every now and then. Remarkably, some animals actually cross the Columbia Icefield during the winter – skiers and climbers occasionally see the tracks of wolverines, goats or mountain caribou on the snow-covered ice.

Smaller mammals like the Columbian ground squirrel are numerous even above treeline. Gray jays, Clark's nutcrackers and ravens are common birds in the area of the ice. So are ptarmigan, ground dwelling grouse-like birds that hide in the tundra meadows surrounding the Columbia

Mountain Goat

(Top) A distant relative of the antelope, the mountain goat is especially well adapted to year-round life in the highest altitudes of the Rockies. Its hooves have cup-shaped pads that secure better footing on rock. Goats can sometimes be seen in the lower valleys when they visit natural salt licks.

Grizzly Bear

(Bottom) The high altitude meadows surrounding the Columbia Icefield are good bear habitat. Both black and grizzly bears can be found in the area. The grizzly is larger than the black bear, and is usually brown or cinnamon in colour, with a hump on its shoulders and a dish-shaped nose.

Bighorn Sheep
Herds of Rocky Mountain Bighorn Sheep frequent the Columbia Icefield area. They are often seen licking salt from the surface of the highway. Salts are usually absent from their diet during the long winters and this is one way to replace them. Both male and female sheep have horns. The horns of the females are short and slender. The males, however, establish herd dominance through head-on butting during the mating season. Their horns can be huge.

Icefield. The capacity of these birds to blend into the vegetation in which they live is legendary – when you come upon them they simply seem to rise out of the flowers and the rocks. Where a moment before, there seemed to be only tundra, now there is a bird; and, maybe, if you look carefully, a whole brood of chicks.

The same warm air currents that bring rain and snow from British Columbia bring thousands of insects with them in summer. As they rise with the wind, the temperature makes them torpid and they fall to the ice. These insects become food for rosy finches that make their nests on the rock faces above the ice. Still, not all the birds that fly over the glaciers are so lucky –travellers on the summer glaciers sometimes find the bodies of birds who have succumbed to the cold and died on the ice. More rarely, they find the body of a moose or a wolverine. Picked apart by ravens, these slowly decomposing bodies occasionally even surface from the snouts of the glaciers.

Despite the bitter cold and short summer of the icefield, some life forms are actually capable of surviving in the snow itself. The most common of these is a very ancient and primitive form of algae that grows near the surface on late summer snow, rich in wind-blown organic debris. In full bloom, this algae can turn the snow pink in colour, and some people call it watermelon snow.

Since the days of the Klondike

gold rush, when Robert Service wrote "The Ballad of the Iceworm Cocktail," there has been a great deal of confusion about whether or not iceworms exist here, or even exist at all. The question was answered during the summer of 1913 by Edmond Murton Walker, who taught biology at the University of Toronto and later founded the extensive invertebrate collection of the Royal Ontario Museum in Toronto. Murton was on a field trip to Banff and its famous hot springs when he and his associate T. B. Kurata unexpectedly discovered "icebugs" in the late spring snow. This discovery resulted in the naming of a new order of insects, the Grylloblatta, and the realization that iceworms do, in fact, exist. Though certainly nowhere near the spaghetti-sized specimens in Service's poem, iceworms have been discovered in the snows of Canada's mountain national parks.

Columbian Ground Squirrel (Above left) An animal most summer visitors to the Columbia Icefield will see is the Columbian Ground Squirrel. This small rodent can be found in the tundra meadows near treeline, within view of the Athabasca Glacier. A relative of the gopher, this ground squirrel is highly social. It thrives on alpine vegetation during the brief icefield summer and then often hibernates for as long as nine months in order to survive the long, bitter winter. Please do not feed or bother these curious creatures.

Ptarmigan (Above right) Ptarmigan are ground-dwelling birds related to the grouse family. They use adaptive coloration to hide from predators, and they are so good at hiding in the alpine foliage that visitors have nearly sat on them before discovering their presence. In winter, their plumage turns white.

Of Ice and Men: The Discovery of the Columbia Icefield

Because of the utter isolation of the Columbia Icefield, it was one of the last major geographical features to be discovered in the Canadian West. As is so often the case, the expedition that discovered the Columbia Icefield was actually looking for something else. In April of 1827, a botanist named David Douglas crossed Athabasca Pass, a major trading route through the Rockies north of the Columbia Icefield. While at the summit of the pass, Douglas climbed one of the adjacent mountains and went home to claim that its peak, and the peak next to it, were in the area of 18,000 feet in height. Subsequent explorers, however, could find no signs of these Himalayan-sized peaks wherever they looked.

Professor Arthur P. Coleman, a geologist with the University of Toronto, first came to the Canadian Rockies in the summer of 1884. At that time there was still considerable controversy about the journals of David Douglas's journals. Douglas claimed to have climbed a 6000 metre (19,680 foot) peak near the summit of the pass in April of 1827. Coleman scoured the Great Divide from Banff to what is now Jasper in search of Douglas's giants. Though he failed to find mountains of that scale, Coleman did discover the tranquil beauty of what would later become the route of the Icefield Parkway.

One explorer in particular, the British mountaineer Norman Collie, was interested in making the first ascent of one of David Douglas's giant peaks. His first challenge, however, lay in finding them. In July of 1898, Collie and his friends Hugh Stutfield and Herman Woolley departed Laggan, near Lake Louise, with a large pack string organized by the famous Banff outfitter Bill Peyto. After nearly three weeks on the trail and many delays caused by high water and rough country, the party found themselves near the summit of the divide separating the Saskatchewan and Athabasca-Sunwapta drainages. On the morning of August 18th, Collie and Woolley started for the summit of Mt. Athabasca, while Stutfield went hunting to see if he could supply the expedition with enough food to stay in the area for a few more days.

Soon Collie and Woolley were on the east side of the peak and climbing. When the ridge gave way to rotten rock, they moved up the glacier and from the northeast ridge made their way to the summit. It was from this incredible vantage point that they discovered an icefield extending to almost every horizon. Collie later reported what they saw:

The view that lay before us in the evening light was one that does not often fall to the lot of modern mountaineers. A new world was spread at our feet: to the westward stretched a vast ice-field probably never before seen by the human

A.P. Coleman

(Top right) Arthur Coleman, a Canadian-born mountaineer and explorer, was a Professor of Geology at the University of Toronto. He, too, had attempted to solve the problem created by David Douglas's claim that mountains nearly twice as high as anything else in the Rockies existed in what is now Jasper National Park. He bypassed the Columbia Icefield in his exploration of the surrounding area in 1892. He was, however, the first to document the pink snow of the Rockies, to which he ascribed the latin name protococcus novalis, or snow algae.

Norman Collie

(Middle right) Norman Collie was a famous British chemist who was best known in his professional field as the inventor of neon. He was also a passionate mountaineer who made some very important first ascents and discoveries in the Canadian Rockies. While making the first ascent of Mt. Athabasca in the summer of 1898, Collie and his companion Herman Woolley were the first to see the Columbia Icefield. He described it this way: "The view that lay before us in the evening light was one that does not often fall to the lot of modern mountaineers. A new world was spread at our feet; to the westward stretched a vast ice-field probably never before seen by the human eye, and surrounded by entirely unknown, unnamed, and unclimbed peaks."

James Outram

(Bottom right) James Outram suffered a nervous breakdown which caused him to abandon his calling as a minister. This development turned out to be the basis of Outram's great fame as a mountaineer. Outram and the Swiss guide Christian Kaufmann made first ascents of some of the most remote and difficult mountains in the Rockies, including Mt. Columbia which the pair climbed in the summer of 1902.

David Douglas
David Douglas was a Scottish botanist who passed through the Rockies north of the Columbia Icefield in the spring of 1827. In his account of the journey, he described Himalayan-sized mountains along the Great Divide of the Rockies. Norman Collie and Hugh Stutfield were looking for these mountains when they discovered the Columbia Icefield in 1898.

eye, and surrounded by entirely unknown, unnamed and unclimbed peaks.

Though some of the peaks he saw were immense and impressive, Collie saw nothing of the scale described by David Douglas. While the heights of the highest peaks of the Canadian Rockies were thus reduced somewhat in scale, the discovery of the Columbia Icefield added greatly to the glamour of these remote mountains. It wouldn't be long before others would follow Collie and Peyto's route to the big glaciers. Though it would be 37 years before all the giants Collie had discovered would finally be climbed, the alpine Olympics had officially begun.

James Outram was the eldest son of a British Baronet. After ten years of active religious service, Outram had a nervous breakdown which forced his early retirement. In 1900, he came to the Canadian Rockies to recover his health. In 1901, in utmost secrecy, Outram

stunned the North American mountaineering community by capturing the first ascent of Mt. Assiniboine, the Matterhorn of the Rockies. A year later, Outram made ten first ascents of peaks over 3050 metres (10,000 feet) and surveyed four new mountain passes in the Columbia Icefield area. One achievement of his 1902 climbing season was the first ascent of Mt. Columbia. Another was the spectacular first ascent of Mt. Bryce, one of the most remote and difficult summits in all the Rockies.

Outram's daring was not duplicated until after the First World War, when Americans James Munroe Thorington, W.S. Ladd and the famous Austrian guide Conrad Kain set a new endurance record for Canadian mountaineering. Driven by sheer enthusiasm, the trio climbed the difficult North Twin, Mt. Saskatchewan and Mt. Columbia in five days.

Amazingly, their Herculean feat was rivalled the very next year by yet another American expedition, this time led by William O. Field. In the company of another famous guide, Edward Feuz, the party climbed both the North and South Twins in just over 24 hours, a combined distance of nearly 60 kilometres (37 miles). Why the rush? The weather in the Columbia Icefield area is highly unpredictable and good conditions seldom last. In 1927, A.J. Ostheimer and two companions put a new route up the North Twin, made the first ascents of Mt. Stutfield

Early Snow Machine Travel on the Columbia Icefield

The Icefields Parkway was a depression relief project that was completed in the fall of 1939. Though the war stalled tourism, it did bring military attention to the Columbia Icefield as an accessible training area for mountain troops. Army half-track vehicles were operated on the Athabasca Glacier during World War II prompting one Jasper entrepreneur to consider a commercial operation on the ice. These pictures show what that operation looked like in the early 1950s. The top photograph shows the chalet as it was then, with one wing, the road to the toe of the glacier, and the ice road up the to where the icefalls could be seen in the distance. The bottom photograph shows early tracked vehicles on the Athabasca Glacier.

and Mt. Kitchener, and made the first traverse of Snow Dome in only 36 hours. These were ambitious adventurers in extraordinary physical condition. During the 63 days they spent in the Rockies, the trio walked over a thousand kilometres (620 miles) and climbed 30 peaks. Twenty-five of the summits were first ascents.

In March of 1932, Cliff White, Joe Weiss and Russell Bennet skied from Jasper to Banff, on a journey of nearly 500 kilometres (310 miles). In the Columbia

Icefield area they made what for years remained the longest continuous ski run in the history of Canadian skiing. Climbing to the summit of Snow Dome, they made a descent of nearly 3000 metres (9840 feet) in a downhill run that lasted for a full 50 kilometres (31 miles). It was runs such as these that awakened the world to the ski potential of the Canadian Rockies.

Six Easy Ways To Experience the Columbia Icefield

You don't have to be an expert mountaineer or ice climber to experience the Icefield.
1. From the Icefield Centre, you can walk an easy 1 kilometre (.62 mile) loop through the forefield of the Athabasca Glacier.
2. For a more vigorous walk, try the hike up Parker's Ridge out of Banff National Park for a view of Saskatchewan Glacier. The trail is 2.4 kilometres (1.5 miles) long with the view at the end.
3. The best day hike with a view of the Icefield is Wilcox Pass. Start from Wilcox Campground, about 2.5 kilometres south of the Icefield Centre. The trail is 6.5 kilometres (4 miles) through alpine meadows, with an elevation gain of 390 metres (1279 feet), and a grand view of the Icefield across a valley.
4. Icewalks on Athabasca Glacier with an expert guide can be arranged through the Icefield Centre. This is the safest way to actually walk on the ice.
5. Snowcoach rides on Athabasca Glacier are safe, fun, and another way to get really close to the ice.

Mountaineers and skiers still come to the Icefield to recreate the famous early climbs. The most popular of these ascents is still Collie's classic route up Mt. Athabasca. Though the traditional route up Mt. Athabasca does not follow exactly the one that Collie and Woolley took in 1898, it follows it closely enough to expose the climber to what these pioneers did and saw. Though absolutely not recommended except to climbers with considerable ice and snow experience, seasoned mountaineers often make this ascent simply for the extraordinary view it offers of the Columbia Icefield. Professional guides can be hired in Jasper for the climb and detailed information about the route and its condition is available in summer at the Icefield or from the Jasper National Park Warden Service. Climbers must register with the Canadian Parks Service before and after attempting this immense and historically significant peak.

Vantage: Parker Ridge

A popular way to explore the Columbia Icefield is to take the three kilometre (1.8 mile) trail up the rolling shoulders of Parker Ridge. This beautiful ridge is named for well-known American climber Herschel Parker, who visited the Columbia Icefield area with Walter Wilcox in 1896. The wide trail up the ridge wanders through old firs and Englemann spruce trees at the edge of timberline, and into the open alpine. In summer it is a natural rock garden, re-splendent with every imaginable colour of wildflower. Before cresting the ridge, the trail also passes impressive 350-400 million year old fossil corals. Once the summit is reached, the Saskatchewan Glacier and its outwash plain dominate the view. Nearly twice as long as the Athabasca Glacier, the Saskatchewan flows gently from the high cold of the icefield into a deeply cut valley that falls steeply off from the viewpoint at the end of the trail.

Though this is one of the very best short hiking trails in all of the Canadian Rockies, it has been badly abused by inexperienced hikers who won't stay on the trails in springtime and by those who destroy the fragile tundra by short-cutting on their return trip down the gentle switchbacks. If you do not have footwear that will allow you to stay strictly on the trail, you really shouldn't attempt this walk. If you do have proper shoes, though, and you are prepared to make a few concessions to the fragility of alpine vegetation, a journey to the top of this ridge could be a turning point in your appreciation for the Icefield.

For More Information

Columbia Icefield:
A Solitude of Ice
Don Harmon and Bart Robinson
Altitude Publishing and The Mountaineers, 1981
This beautiful book combines the outstanding photography of Banff photographer Don Harmon with a sometimes haunting text written by *Equinox* editor Bart Robinson. It is a good introduction to human reactions to the overwhelming beauty of the Columbia Icefield since its discovery in 1898.

How the Experts Experience the Columbia Icefield

The Parks Service offers a registration service for those undertaking hazardous activities. If you don't register out by your estimated time, a search and rescue team will come looking for you. Overnight camping in the parks requires a backcountry permit.

1. Mount Columbia is the most spectacular hike around the Icefield, and at 3747 metres (12,290 feet), it's also the tallest peak in the area. This is a very technical scramble.

2. The 3491 metre (11,450 foot) Mount Athabasca is another very technical climb with a great view.

3. Backpacking is not as technically demanding as slogging through snow and ice up a mountain, but backpacking in this area is not for amateurs. The Skyline Trail in Jasper National Park is 44.1 kilometres (27 miles) long, with views of the Columbia Icefield, Athabasca River, and the Maligne Lake peaks. The hike usually takes two or three days.

4. Ski touring on the Icefield is another option for those with enough experience to do it safely.

5. Frozen waterfall climbing is breathtaking to watch, incredibly difficult to master, and perhaps the most specialized way to get close to the ice. A spectacular spot near the Icefield is the Weeping Wall in the northern end of Banff National Park, just off the Icefields Parkway.

The Rocky Mountains of Canada North
Robert Kruzyna and William L. Putnam
American Alpine Club, Alpine Club of Canada Climber's Guide, 1985
This is the bible of mountaineering routes for climbers visiting the Canadian Rockies north of Lake Louise. It offers an extensive examination of climbing opportunities in the Columbia Icefield area.

Exploring the Columbia Icefield
Richard E. Kucera
High Country Publications, 1981
This small booklet, written by one of Canada's leading experts in glacial dynamics, offers an accurate, straightforward analysis of the features created by advancing ice in the Athabasca Glacier area.

The Canadian Rockies SuperGuide
Graham Pole
Altitude Publishing, 1991
This is one of the most readable and engaging all-purpose guides available on the natural and human history of the Canadian Rockies. It also offers excellent information on the Icefield Parkway.

Glacier Ice
Austin Post and Edward R. Lachapelle
University of Washington Press, 1971
Though it may be a little hard to get now, this large-format book is a classic. It is one of the few books on this subject that takes a comparative look at glacial features all over the world in order to come to some conclusions about the climates and circumstances that create them.

Glossary

alpine The life zone in mountainous regions that extends above treeline. In the Canadian Rockies, depending on slope and exposure, treeline ranges in altitude, from approximately 2100 to 2250 metres (roughly 6800 to 7200 feet) above sea-level. Life does extend above the trees, usually in the form of high meadows composed of inter-connected rafts of plant life. The alpine in the Canadian Rockies is also referred to as alpine tundra, for it is, in many ways, a southward extension of the arctic tundra climate and growing conditions in the polar regions.

alpine valley glacier A small mountain glacier fed by accumulation of snow in the same trough down which it flows.

bergschrund A large semi-permanent crevasse at the head of a glacier which separates flowing ice from stagnant ice or from rock above.

Carboniferous A period of the Palaeozoic Era in the Earth's history. It occurred between the Devonian and Permian periods, from approximately 400 million to 225 million years ago. During this period, coal beds were laid down as central features of the stratigraphy of many parts of the north hemisphere.

Cenozoic The most recent geological era in the Earth's history, extending from roughly 75 million years ago to the present. This geological era includes the Tertiary and the Quaternary periods. Much of the glaciation that has shaped the Canadian Rockies as we know them today has occurred during this era.

chlorofluorocarbons Commonly known as CFC's, this group of aerosol propellants and refrigerants was initially developed by American scientist Thomas Midgley in 1930. In 1974, it was discovered that these hardy compounds remain in the lower atmosphere for 75 to 120 years, where they break down into chlorine atoms, each of which is capable of destroying up to a hundred thousand ozone molecules before falling back to Earth. These chemicals cause huge holes in the Earth's atmosphere, endangering all living things on the planet.

continental divide A divide is a height of land separating two different watersheds. A continental divide separates watersheds that flow into different oceans. In the case of the Canadian Rockies, the continental divide, or Great Divide as it is often called, separates waters flowing from the Pacific Ocean from those that flow to the Atlantic. At the Columbia Icefield, however, there is a triple continental divide, which is the high dividing point that separates waters flowing into the Atlantic, Pacific and Arctic oceans.

Cretaceous A geological period in the history of the Earth that extended from roughly 150 to 75 million years ago, and was marked by the wholesale disappearance of many of the planet's earlier life forms.

crevasse A crack or fissure in glacial ice. The crevasse is a break in ice, whereas a crevice is a narrow opening resulting from a split or crack in rock.

firn line The often quite obvious boundary on the upper stretches of the glacier, above which snow does not melt. The firn line distinguishes where a glacier ends and the icefield that forms it begins.

geology The science that deals with the history of the Earth as recorded in fossils found in the rocks of its crust.

glacier A major body of ice that moves under the influence of gravity.

graupel Granular snow pellets often called soft hail.

Great Glaciation A major glacial period that began in North America roughly 240,000 years ago. This major geological event appears to have lasted a hundred thousand years.

hoar A sparkling, crystalline form of frost found on, above or below the snow's surface. Depth hoar is recrystallized snow found in the bottom layers of a snowpack.

hypothermia The cooling of the core of the body to sub-normal temperatures.

icecap A glacier forming on an extensive area of relatively level land and flowing outwards from its centre. A mountain ice cap is a flat or gently sloping alpine upland buried in ice.

icefield An upland area of ice that feeds two or more glaciers.

iceworm A small annelid worm that spends its entire life cycle on snow or ice.

iridium A very heavy metallic element of the platinum group that is silver-white, hard and brittle.

Jurassic A geological period in the Earth's history, well-represented in fossil records by abundant dinosaur life and the introduction of the first mammals.

Little Ice Age A localized glacial advance in the Canadian Rockies that appears to have began somewhere around 1200 A.D., peaked in around 1750 and ended around 1900.

millwell A vertical hole by which a surface meltwater stream enters a glacier. A millwell is the same as a *moulin* or a glacier mill.

moraine A deposit of rock debris shaped by glacial flow and erosion. Several types of moraines exist in the area of the Columbia Icefield, including terminal, lateral, medial and ablation moraines, each of which is formed by different types of glacial action.

moulin See *millwell*.

ogive A band or undulation of dark and light ice on the surface of a glacier, usually recurring in a periodic pattern.

Ordovician A geological period in the Earth's history roughly between 500 and 450 million years ago. Fossil fishes have been discovered from this period and, on land, the preserved remains of mosses.

outlet valley glacier A glacier that flows out of a major icefield accumulation zone and into a neighbouring valley.

Permian A geological period in the Earth's history, roughly between 275 million and 225 million years ago. It is represented in fossil records by great cycads and conifer forests.

pink snow Snow made pink by the presence within it of a form of algae that survives on organic detritus that blows onto old snow in summer.

Precambrian One of the earliest geologic periods in the history of the Earth. The only biological records from this period, which ended roughly 620 million years ago, include invertebrate animals, isolated plant spores and the fossils of ancient marine algae.

rock flour Rock that has been ground into a fine powder by glacial ice. This fine debris gives a milky colour to rivers and a light brown or green colour to lakes fed by glaciers. The suspension of rock flour in many of the lakes in the area of the Columbia Icefield give the water the colour of cobalt. The dazzling blue of Peyto Lake and Lake Louise as seen from above is caused by suspended rock flour.

serac A standing tower of ice breaking off a glacier to form an icefall.

solar wind The great storm of light and radioactive particles through which the Earth spins as it passes through the sun's glow.

snowcoach A 56-passenger vehicle developed specifically for travel on the Athabasca Glacier at the Columbia Icefield. (Also spelled "snocoach.")

Triassic That period in the history of the Earth from approximately 225 to 200 million years ago, represented in the strata of this planet's surface by abundant land and sea life.

watermelon snow See *pink snow*.

watershed The area drained by a river system, or a ridge dividing the areas drained by different river systems.

An award-winning photographer and film-maker, Robert Sandford is also the author of five books on the natural and human history of the Canadian Rockies, three of which have been published by Altitude. Mr. Sandford and his family live in Canmore, Alberta, just outside of Banff National Park.